# Macramè Mini-guide for Beginners

## How to Macramè, Knots Boot, Projects and Instructions

Karen Fisher

© Copyright 2021 by Karen Fisher- All rights reserved.

The following Book is reproduced below to provide information that is as accurate and reliable as possible. Regardless, purchasing this Book can be seen as consent because both the Publisher and the author of this Book are in no way experts on the topics discussed within. Any recommendations or suggestions that are made herein are for entertainment purposes only. Professionals should be consulted as needed before undertaking any of the actions endorsed herein.

This declaration is deemed fair and valid by both the American Bar Association and the Publishers Association Committee. It is legally binding throughout the United States.

Furthermore, the transmission, duplication, or reproduction of any of the following work, including specific information, will be considered an illegal act irrespective of if it is done electronically or in print. This extends to creating a secondary or tertiary copy of the work or a recorded copy and is only allowed with the Publisher's express written consent. All additional rights reserved.

The information in the following pages is broadly considered a truthful and accurate account of facts. As such, any inattention, use, or misuse of the information in question by the reader will render any resulting actions solely under their purview. There are no scenarios in which the Publisher or the original author of this work can be in any fashion deemed liable for any hardship or damages that may befall them after undertaking the information described herein.

Additionally, the following pages' information is intended only for informational purposes and should thus be thought of as universal. As befitting its nature, it is presented without assurance regarding its prolonged validity or interim quality. Trademarks that are mentioned are done without written consent and can in no way be considered an endorsement from the trademark holder.

# Table of Contents

**INTRODUCTION** ................................................................................. 5

**MACRAMÉ FOR BEGINNERS** ........................................................ 8

    MAIN KNOTS USED ............................................................................ 15
    CAVANDOLI MACRAMÉ ..................................................................... 17
    IMPORTANT MACRAMÉ TERMS ........................................................ 17
    COMMON MATERIALS USED ............................................................ 32
    BEGINNER'S BRACELET .................................................................... 34

**MACRAMÉ KNOTS BOOK** ............................................................ 36

    ALTERNATING SQUARE KNOTS ....................................................... 36
    CAPUCHIN KNOT ............................................................................... 43
    CROWN KNOT .................................................................................... 46
    DIAGONAL DOUBLE HALF KNOT .................................................... 50
    FRIVOLITY KNOT ............................................................................... 53
    HORIZONTAL DOUBLE HALF KNOT ............................................... 57
    JOSEPHINE KNOT ............................................................................. 59
    LARK'S HEAD KNOT ......................................................................... 62
    REVERSE LARK'S HEAD KNOT ........................................................ 65
    LARK'S HEAD HALF STITCHES KNOT ............................................. 68
    SINGLE HALF STITCH ....................................................................... 71
    SPIRAL STITCHES ............................................................................. 74

**MACRAMÉ PROJECTS I** ............................................................... 75

    DIY MACRAMÉ FEATHERS ............................................................... 75
    DIY TASSEL AND MACRAMÉ KEYCHAINS ..................................... 77
    SIMPLE MODERN DIY MACRAMÉ WALL HANGING ..................... 78
    DIY ROUND MACRAMÉ BOHO COASTERS ..................................... 81
    MACRAMÉ CURTAIN ......................................................................... 82
    FRIENDSHIP BRACELET WATCH ...................................................... 85

**MACRAMÉ PROJECTS II** .............................................................. 88

    SERENITY BRACELET ....................................................................... 88
    LANTERN BRACELET ....................................................................... 94
    CELTIC CHOKER ............................................................................... 98
    CLIMBING VINE KEYCHAIN ............................................................ 104
    FILIGREE LANCELET BRACELET .................................................... 109

**INSTRUCTION BOOKS** .......................................................................... **114**
    PLANT HANGER AYLA .................................................................115
    PLANT HANGER BELLA ...............................................................121
    PLANT HANGER CATHY ..............................................................129
**CONCLUSION** ..................................................................................... **137**

# Introduction

Macramé is a Type of textile-making that doesn't demand the conventional method of weaving or knitting, but alternatively by way of a set of knots. It's deemed to have begun since the 13th century in the western hemisphere with all the Arab weavers. They'd not get the surplus ribbons and yarn onto the endings of both hand-loomed cloths such as towels, veils, veils, and shawls into decorative fringes. We discovered it fascinating that sailors had been people to genuinely create this alluring and have been imputed to dispersing this art to various states throughout the vents they'd stop. They'd decorate the handles of knives, bottles, and other things that can be discovered on the boat and utilize them to find something they wanted or wanted whenever they reached land.

Regarding this, nineteenth-century sailors generated hammocks and straps, having an activity referred to as "square foot." Materials that Are Often utilized for Macramé are Cotton twine, hemp, yarn, or leather. While there are variations, the principal knots would be the square Knot, although complete feasibility and double half hitches. Jewelry is usually developed by blending ribbons with diamonds, diamonds, rings, diamonds, or cubes. In the event you have a look at the vast bulk of the friendship outfits exhausted by faculty kiddies, you will learn they will have been created through the use of Macramé.

After analyzing these basic Knots, which are often utilized in creating Macramé, I came around the cavaedia Macramé. This design comprises two colors that consist of 2 main knots that can be left, making a milder sort of fabric that works excellent for dining table mats, purses, publication, etc. Along with covers. Cavaedia Macramé is well known as Valentina cavaedia, who gained a golden trophy of fame out of 1961 before she passed on at age 97 in 1969. In Italy, this exceptional lady became the headmistress of a house into the evil or orphaned kids in turn through the entire conclusion of this first world war. This is a center where approximately 100 youths may be placed between the ages of 1-5. To keep the kiddies busy, she educated them on art she had heard from her great-grandma, Macramé. The kiddies would create matters to advertise in marriage savings, and attentive records were kept of each child's income and might be spread in their mind when they'd leave your house. Regrettably, your house at which she had been casa del only just lived until 1936 when due to this political situation in Italy it was too tricky because of the benefactors of your house to keep on. The enthusiasm for Macramé appeared to fade for Some Time. However, it was widely used from the 1970s by the American neo-hippies and grunge audiences in producing jewelry. This art was comprised of handmade bracelets, anklets, and bracelets adorned with handmade glass beads and natural elements such as shell and bone.

# Macramé for Beginners

There are plenty of places available for people who want to learn how to macramé.

Macramé focuses on creating involved nodes that generate complete designs that can be turned into elegant bracelets, flower pots, and decorative wall hangings.

If you are interested in this topic, the first and least complicated step in learning how to macramé is to understand the basic knots and diagrams. If you want to add some macramé to your furniture, you can choose to pursue a DIY project or purchase something similar in a home products shop. For DIYers, start with a simple project to avoid getting frustrated and stopping before you finish. Mastering the superficial nodes is an important step that allows you to find more advanced knotting techniques more easily.

Try using items you already have before buying a whole stock of macramé supplies for your first project. You will want to be sure that you love this sport before you invest much money. Choose one thread type and one assembly ring. Search around the house for a work surface frame. You can purchase those macramé pins to connect the file to the wall. Nonetheless, for your first project, you can use safety pins to save some money.

The online network is an excellent place to start looking for alternatives to macramé.

Visual aids are of enormous help and make learning how to macramé easy.

For many people, it is much easier to follow diagrams rather than written directions, which are hard to comprehend.

And once you have become familiar with the visual aids, it is time to get the supplies to start the macramé process.

A diagram cannot help your macramé correctly, irrespective of its thoroughness and how well it is explained.

There is a need for a thread that can effectively macramé efficiently.

Like any acquired art, it is practical to try to learn how to macramé.

Get some straightforward, basic diagrams to get training representations going.

The simpler ones are less complicated than the complexities of the sophisticated ones.

You will lead them with plenty of time and exercise. Apart from simple knot instructions, you will also have to study and practice a little before you can memorize and make comfortable knots.

That's not going to be taught if you rush, step by step, to learn how to macramé. Once you have mastered the simple knot patterns, line them up to do straightforward works like bracelets. In addition to the ties, you also need an eye to match the best colors to produce the knotwork. Bracelets are ideal for beginners, as the most comfortable knots without difficulty are necessary. You can work with very complex trends when you feel more confident with your abilities. The biggest positive about intricate and too complex structures is that they can be designed solely to create exceptional ornamental objects. Determining how long you need to learn how to macramé depends on many factors, such as how quickly you can learn the technique. Suppose you have knitted or sewn for a long time. In that case, the degree of difficulty should be considerably lower because the process is identical. Macramé is a fun craft to learn, and with a small budget, you can continue. There are plenty of free or fair trends and a great way to start reading. This is a beautiful art to include your family, grandchildren, or anyone.

## Macramé is Everywhere

You don't need to look for a yoga studio to teach you how to knit a knotty board, a drapery plant hanger, or a small key chain together. Boho and minimalist variations provide a wet yet trendy feel. Macramé supplies-You want to get the materials going!

Macramé materials are typical for cording, ringing, pins, work boards, and beads. Such main supply types include various scale, form, and content combinations. Various ventures, together with expectations, will determine the exact types of macramé provisions.

There are two primary forms of macramé cords: **Natural and Synthetic**. Jute is typically a natural, twin-like fiber, whereas synthetic macramé cords appear to be smooth. These can be ordered in a range of colors. Synthetic macramé cord is often bent instead of twisted in its shape.

The rings for macramé designs vary in size from keyring to large hoops. Often, hoops or rings can be used as macramé frames connected with the knots still fixed. T-pins are macramé materials used to keep the work on aboard. Macramé boards are often triangular, compact, and made of wood or cork pressed. T-pins join the boards to start a macramé project and encourage you to work peacefully on your project without risking losing your seat or action, which could be detrimental to the outcome. Such pins are known for their broad cross-section, which allows maneuverable pins. Many people find macramé that spinning loops of cords vertically with a board is less complicated and less likely to tangle their designs than on horizontal surfaces. In many macramé bits, beads are used. These are available in wood, ceramic, and plastic. Round, oval, and cylindrical are popular forms of beads for macramé supplies. Wood beads could be colored light or black. The plastic beads used in macramé designs could be translucent or opaque. In contrast, ceramic macramé beads, such as painted floral motifs, appear to be framed.

**Macramé supplies-What to start with!**

The principal thing you need from your macramé jewelry supply is the cord for knotting designs. One of the most common is hemp which is a lock or twine from a hemp plant. It is super strong and long-lasting. It is now possible to achieve in a wide range of colors and the old "styles. "You only need a few more things to continue after you get your chain. The surface area you are working on is the most critical resource. I started with a clipboard and put my cords on a pencil. A foam pillow that I covered with fabric is my personal favorite. A piece of corkboard is another choice. These can be found in most craft stores. I recommend it is at least 3/8 inches thicker or thicker and about 11x17 inches thicker. You need a big enough section to be put conveniently on your lap to lean against a wall. You will need some lovely strong pins to hold your project to your work area. I propose either sewing pins or T-pins to keep wigs on the heads of foam. If your string is fragile as satin thread, the sewing pins should be used up with colored balls. You won't leave a massive hole like sometimes the T-pins. The hole diameter is your only limitation. Most cords are usually about 1m thick.

**Project suggestions:**

You can execute so many different macramé designs. Each layout is composed of hundreds of variations. You will create your patterns and develop some genuinely unique textiles once you become accustomed to knotting.

Think of ways to change a few of these macramé ideas:

**Planters hanging keychains**

**Belts**

**Jewelry fringe on other textiles**

Millennials may have brought macramé back, but people of all ages can appreciate it and fall in love with it. Macramé craze faded over time—but, in the 70s, it became popular once more as a means to create draperies, table cloths, wall hangings, jean shorts, bedspreads, and even plant hangers. During this time, Macramé started being popular as jewelry for the grunge and neo-hippie crowd.

# Main Knots Used

**1. Square/Reef Knot.** This is the primary knot used. This is done by binding the line or rope around a particular object. It is also known as the base knot. You could make it by tying a left hand over a knot over a right hand. In short, *right over left* and *leftover right.*

**2. Half Hitch.** This is done by working the end of one line over the standing part of The Knot. It is one of the most valuable knots, bends, and hitches, among anything else.

**3. Overhand Knot.** Another knot you could use is the overhand knot. It is known as one of the world's fundamental stitches and is especially helpful in Macramé. To tie, you could simply loop a thread to the end with the help of your thumb. Or, you could also twist a bight by placing your hand over your wrist as you loop. Use your fingers to work to the end.

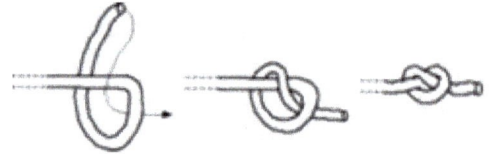

4. **Crown Knot.** Crown knot means that you have to go over the knot, go under twice, over twice, and under again. You'll then create something like what's shown in the illustration below.

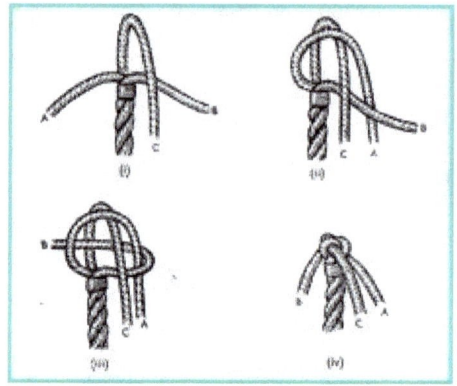

# Cavandoli Macramé

Cavandoli Macramé is the term given to Macramé that allows you to create intricate knots—perfect for jewelry and home décor! It was named after Valentina Cavandoli, who experimented with different knots to create her Macramé patterns while teaching Montessori.
It's done by using both the square and half hitch knots—so you could be as creative as possible.
These different knots then balance each other out, helping you create some of the most beautiful projects around!

## Important Macramé Terms

Of course, you could also expect specific terms you would be dealing with while trying Macramé out. By knowing these terms, it would be easier for you to make Macramé projects. You won't have a hard time, and the crafting would be a breeze!
For this, you could keep the following in mind!

# Alternating

This is applied to patterns where more than one cord is being tied together. It involves switching and looping, just like the

half-hitch.

## Adjacent

These are knots or cords that rest next to one another.

## Alternating Square Knots (ASK)

You'll find this in most Macramé patterns. As the name suggests, it's all about square knots that alternate on a fabric.

## Bar

When a distinct area is raised in the pattern, it means that you have created a "bar." This could either be diagonal, horizontal,

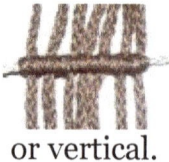

or vertical.

## Bangle

The bangle is the term given to any design with a continuous

pattern.

## Band

A design that has been knotted to be flat or wide.

## Buttonhole (BH)

It's another name given to the Crown or Lark's head knot. It has been used since the Victorian Era.

## Button Knot

This is a knot that is firm and is round.

## Bundle

These are cords that have been grouped as one. They could be held together by a knot.

## Braided Cord

These are materials with individual fibers that are grouped as one. It is also more substantial than most materials because all the fibers work together as one.

**Braid**

Sometimes called Plait, this describes 3 or more cords that have been woven under or over each other.

**Body**

**Bight**

## Crook

This is just the part of the loop that has been curved and is situated near the crossing point.

## Core

This term refers to a group of cords that are running along the center of a knot. They're also called "filling cords."

## Cord

This could either be the material, or cord/thread that you are using, or specific cords that have been designed to work

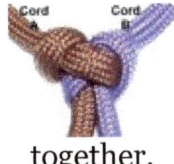

together.

## Combination Knot

These are two knots that have been designed to work as one.

## Cloisonne

A bead with metal filaments that are used for decorative purposes.

## Chinese Crown Knot

This is usually used for Asian-inspired jewelry or décor.

## Charm

This is a small bead that is meant to dangle and is usually just an inch in size.

## Doubled

These are patterns that have been repeated in a single pattern.

## Double Half Hitch (DHH)

This is a specific type of knot that's not used in many crafts, except for really decorative, unusual ones. This is made by making sure that two half hitches are resting beside each

other.

## Diameter

This describes the material's weight based on millimeters.

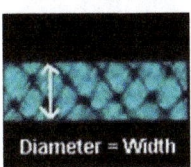

## Diagonal

This is a row of knots or cord that runs from the upper left side to the opposite.

## Excess Material

This describes the part of the thread that's left hanging after you have knotted the fabric. Sometimes, it's hidden using fringes, too.

## Fusion Knots

This starts with a knot so you could make a new design.

## Fringe

This technique allows cords to dangle down with individual fibers that unravel themselves along with the pattern.

## Flax Linen

This is material coming from Linseed Oil that's best used for making jewelry. Even Macramé clothing has been used for over 5000 years already.

## Finishing Knot

This is a kind of knot that allows specific knots to be tied to the cords not to unravel.

## Findings

These are closures for necklaces or other types of jewelry.

## Gemstone Chips

This is the term given to semi-precious stones that are used to decorate or embellish your Macramé projects. The best ones are usually quartz, jade, or turquoise.

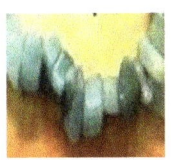

## Horizontal

This is a design of the cord that works from left to right.

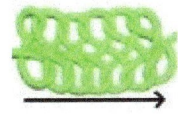

## Holding Cord

This is the cord to where the working cords are attached to.

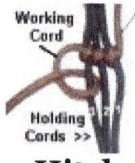

## Hitch

This is used to attach cords to cords, dowels, or rings.

## Inverted

This means that you are working on something "upside-down."

## Interlace

This is a pattern that could be woven or intertwined so different areas could be linked together.

## Micro-Macramé

This is the term given to relatively small Macramé projects.

## Metallic

These are materials that resemble silver, brass, or gold.

## Mount

Mount or Mounting means that you have to attach a cord to a frame, dowel, or ring, and it is usually done at the start of a

project.

## Netting

This is a knotting process that describes knots formed between open rows of space and is usually used in wall hangings, curtains, and hammocks.

## Natural

These are materials made from plants or plant-based materials. Examples include hemp, Jude, and flax.

## Organize

This is another term given to cords that have been collected or grouped as one.

## Picot

These are loops that go through the edge of what you have

knotted.

**Pendant**

A décor that you could add to a necklace or choker and could easily fit through the loops.

**Synthetic**

This means that the material you are using is human-made

and not natural.

**Symmetry**

This means that the knots are balanced.

**Standing End**

This is the end of the cord that you have secured so the knot would be constructed appropriately.

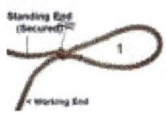

## Texture

This describes how the cord feels like in your hand.

**Soft Texture**

## Tension or Taut

This is the term given to holding cords that have been secured or pulled straight so that they would be tighter than the other working cords.

## Vertical

This describes knots that have been knitted upwards or vertically.

## Working End

This is the part of the cord that is used to construct the knot.

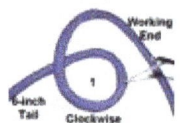

# Weave

This is the process of letting the cords move as you let them pass over several segments in your pattern.

# Common Materials Used

Of course, as you'll be doing various projects, you can expect that you'll also be using various materials. However, there are raw materials that you always have to have with you when creating Macramé projects.

These include the following:

**1. Cords.** The best ones you can use are 3-ply cords made of 3 different fibers twisted together, just like the one shown below. These cords can also be made from yarn, leather, jute, hemp, linen, cotton twine.

3-Ply Cords

**2. Beads.** These are used to create jewelry, such as bracelets, necklaces, and the like. The best beads you can use are those made from glass, wood, or resin.

Example of Macramé Beads

**3. Gemstones.** Sometimes, gemstones are also added to lend more flair to your projects, making them great eye-candy!

Macramé Necklaces made from gemstones

**4. Knotting Board.** Finally, you can use a knotting board to keep your macramé items in place as you create that jewelry!

A Macramé Board that's available in the market.

# Beginner's Bracelet

This is an easy Macramé project that's perfect for beginners!
What you need:
- **Glue-on end clasps**
- **Jewelry glue**
- **Ring connector**
- **Cotton or hemp twine**

### Instructions:

First, take three of the hemp or cotton twine strands and make a loop out of them. Loop the loop that you have made around the connector, and then make a knot on each side.

Now, you'll see 6 strands coming off the sides of your loop after you have inserted it through the connector. Braid each side—you can make simple braids, or even 6-strand braids if you can do it.

Trim the ends off once you get to the end. Make sure to use jewelry glue and secure the braid by gluing it on. Fully twist the end caps to coat the spine of your bracelet. Check the length before securing it so you can be sure that it would fit you.

# Macramé Knots Book

## Alternating Square Knots

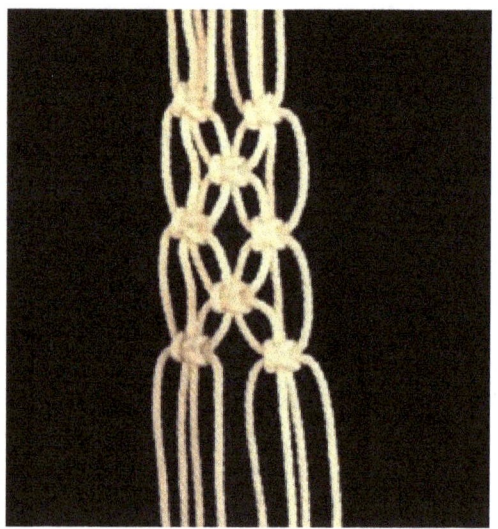

This is the perfect knot to use for basket hangings, decorations, or any projects requiring you to put weight on the project. Use a heavier weight cord for this, which you can find at craft stores or online.

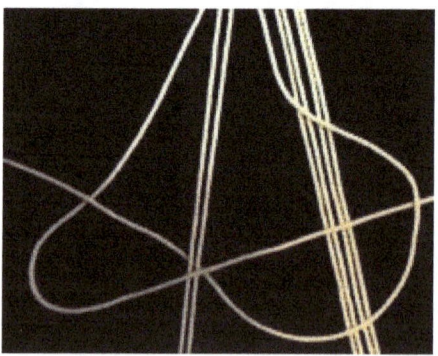

Oversee the photos as you move along with this project, and take your time to make sure you are using the right string at the right point of the project.

Don't rush, and make sure you have even tension throughout. Practice makes perfect, but with the illustrations to help you, you'll find it's not hard at all to create.

Start at the top of the project and work your way toward the bottom. Keep it even as you work your way throughout the piece. Tie the knots at 4-inch intervals, working your way down the entire thing.

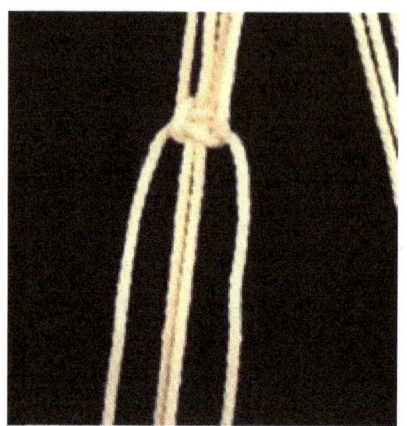

Tie each new knot securely before you move on to the next one. Remember that the more even you get, the better it is.

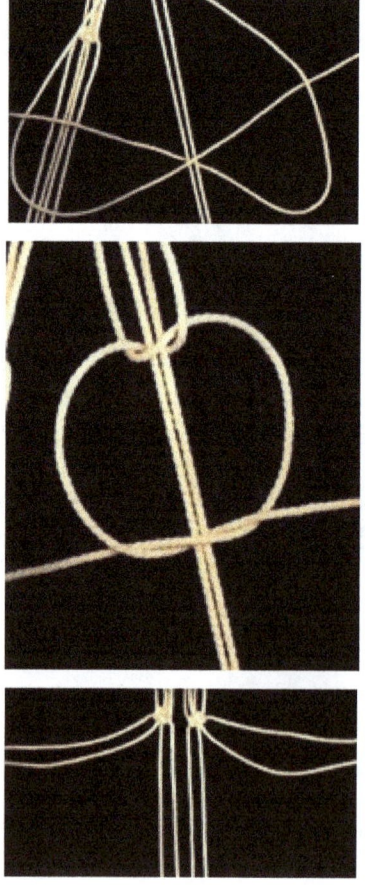

Work on one side of the piece first, then tie the knot on the other side. You will continue to alternate sides, with a knot joining them in the middle, as you can see in the next photo. Again, keep this even as you work throughout.

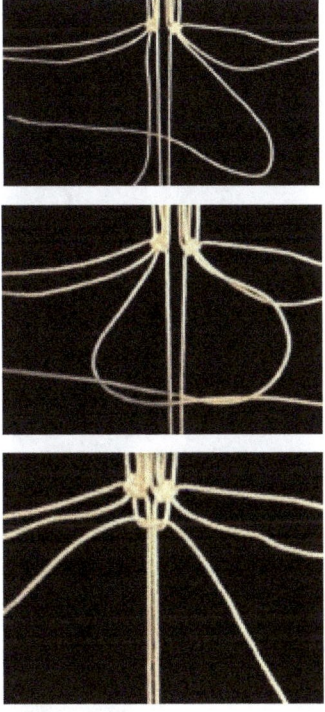

Bring the knot in toward the center, and make sure you have even lengths on both sides of the piece.

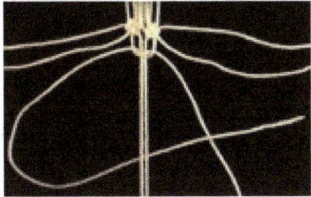

Pull this securely up to the center of the cord, then move on to the next section on the cord.

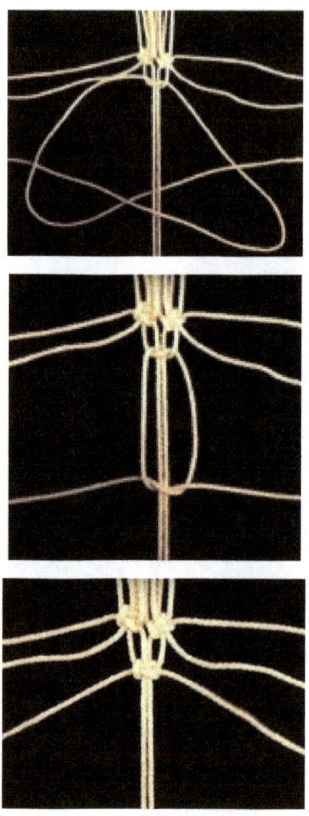

You will gather the cord on one side for the next set of knots; then, you will go back to the other side of the piece to work another set of knots on the other side. Work this evenly; then, you are going to come back to the center.

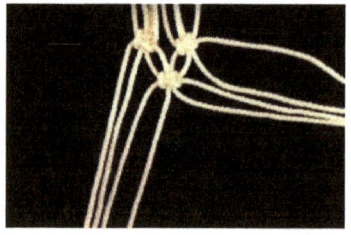

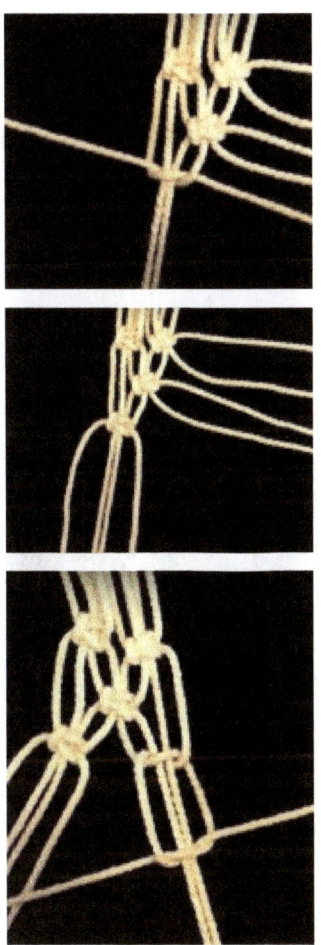

It's a matter of sequence. Work on the one side, then go back to the beginning, then go back to the other side once more. Continue to do this for as long as your cords are or as long as you need for the project.

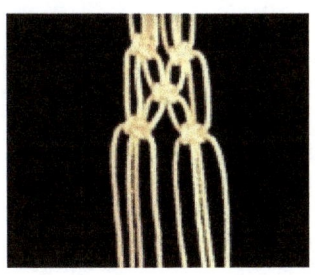

For the finished project, make sure that you have all your knots secure and firm throughout, and do your best to make sure it is all even. It will take practice before you can get it correctly each time, but remember that practice does make perfect, and with time, you will get it without too much trouble.

Make sure all is even and secure, and tie off. Snip off all the loose ends, and you are ready to go!

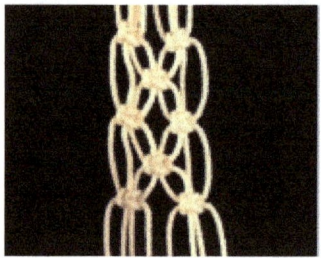

# Capuchin Knot

This is an excellent beginning knot for any project and can be used as the foundation for its base. Use lightweight cord for this – it can be purchased at craft stores or online, wherever you get your macramé supplies.

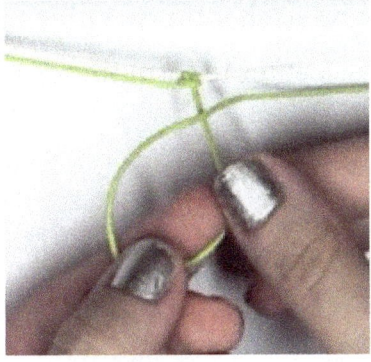

Oversee the photos as you move along with this project, and take your time to make sure you are using the right string at the right point of the project.

Don't rush, and make sure you have even tension throughout. Practice makes perfect, but with the illustrations to help you, you'll find it's not hard at all to create.

Start with the base cord, tying the knot onto this, and working your way along with the project.

Twist the cord around itself 2 times, pulling the string through the center to form the knot.

For the finished project, make sure that you have all your knots secure and firm throughout, and do your best to make sure it is all even. It will take practice before you can get it correctly each time, but remember that practice does make perfect, and with time, you will get it without too much trouble.

Make sure all is even and secure, and tie off. Snip off all the loose ends, and you are ready to go!

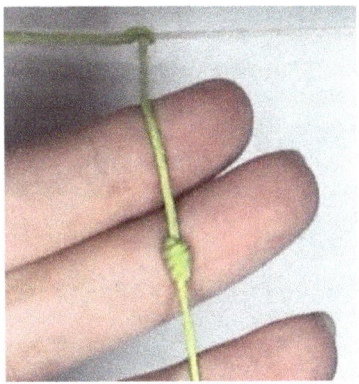

# Crown Knot

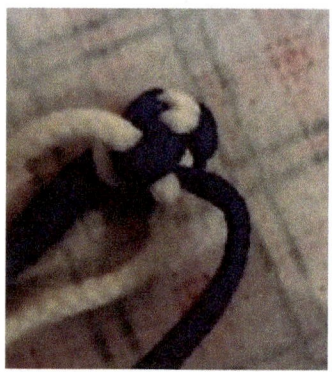

This is an excellent beginning knot for any project and can be used as the foundation for its base. Use lightweight cord for this – it can be purchased at craft stores or online, wherever you get your macramé supplies.

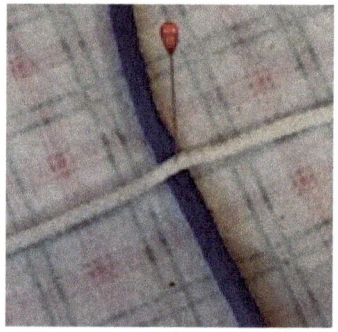

Oversee the photos as you move along with this project, and take your time to make sure you are using the right string at the right point of the project. Don't rush, and make sure you have even tension throughout. Practice makes perfect, but with the illustrations to help you, you'll find it's not hard at all to create.

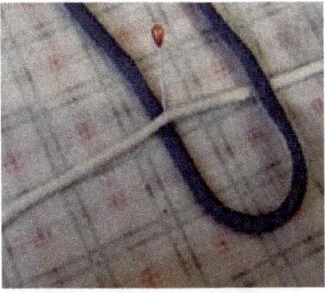

Use a pin to help keep everything in place as you are working.

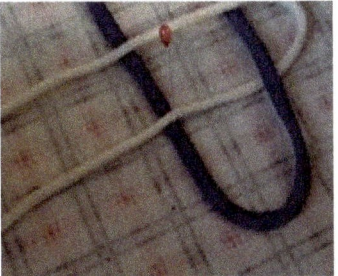

Weave the strings in and out of each other, as you can see in the photos. It helps to practice with different colors to help you see what is going on.

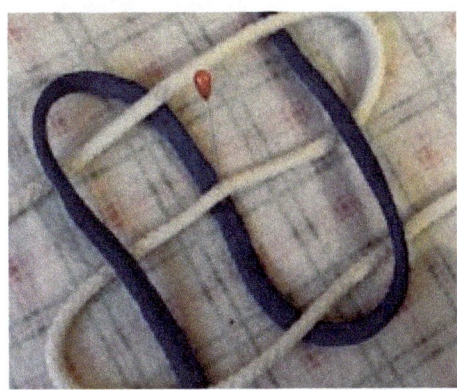

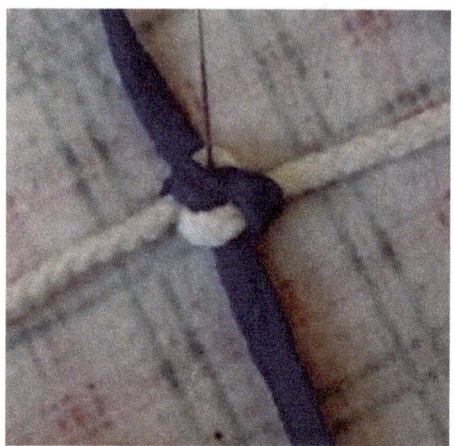

Pull the knot tight, then repeat for the next row on the outside.

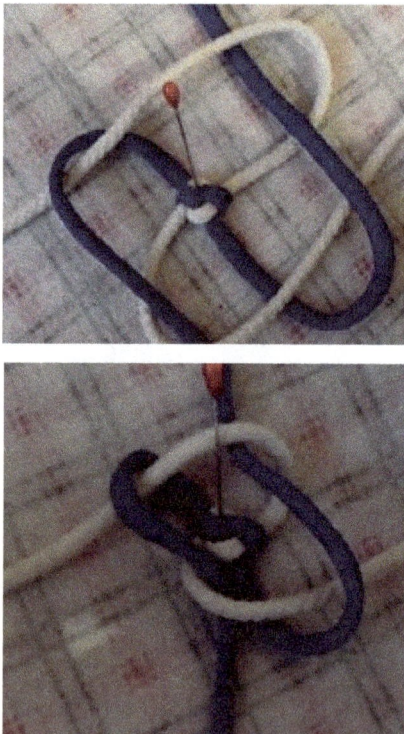

Continue to do this as often as you like to create the knot. You can make it as thick as you like, depending on the project. You can also create more than one length on the same cord.

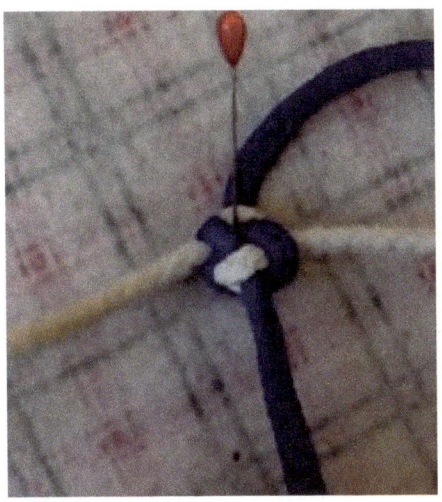

For the finished project, make sure that you have all your knots secure and firm throughout, and do your best to make sure it is all even. It will take practice before you can get it correctly each time, but remember that practice does make perfect, and with time, you will get it without too much trouble. Make sure all is even and secure, and tie off. Snip off all the loose ends, and you are ready to go!

# Diagonal Double Half Knot

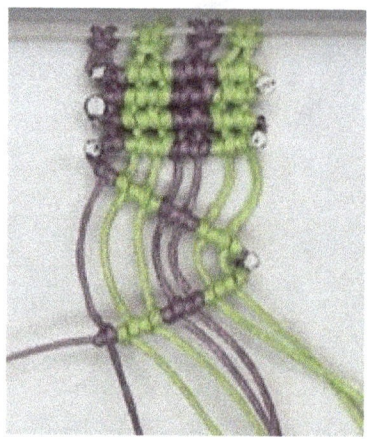

This is the perfect knot to use for basket hangings, decorations, or any projects requiring you to put weight on the project. Use a heavier weight cord for this, which you can find at craft stores or online. Oversee the photos as you move along with this project, and take your time to make sure you are using the right string at the right point of the project. Don't rush, and make sure you have even tension throughout. Practice makes perfect, but with the illustrations to help you, you'll find it's not hard at all to create.

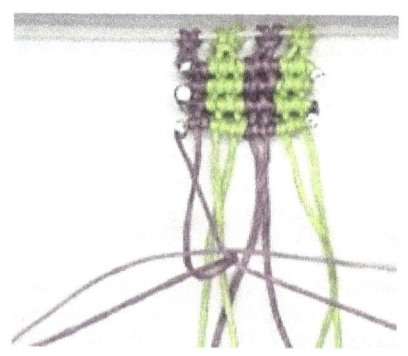

Start at the top of the project and work your way toward the bottom. Keep it even as you work your way throughout the piece. Tie the knots at 4-inch intervals, working your way down the entire thing.

Weave in and out throughout, watching the photo as you can see for the knots' right placement. Again, it helps to practice with different colors to see what you need to do throughout the piece.

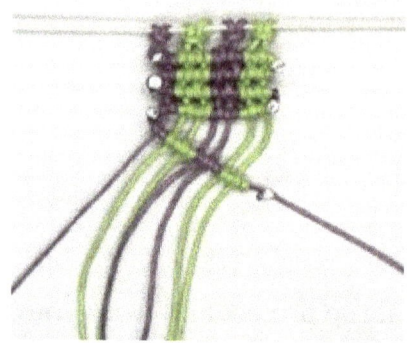

For the finished project, make sure that you have all your knots secure and firm throughout, and do your best to make sure it is all even. It will take practice before you can get it correctly each time, but remember that practice does make perfect, and with time, you will get it without too much trouble. Make sure all is even and secure, and tie off. Snip off all the loose ends, and you are ready to go!

# Frivolity Knot

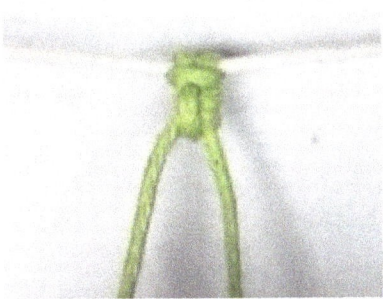

This is an excellent beginning knot for any project and can be used as the foundation for its base. Use a lightweight cord for this. It can be purchased at craft stores or online, wherever you get your macramé supplies.

Oversee the photos as you move along with this project, and take your time to make sure you are using the right string at the right point of the project. Don't rush, and make sure you have even tension throughout. Practice makes perfect, but with the illustrations to help you, you'll find it's not hard at all to create.

Use the base string as the guide to hold it in place, then tie the knot onto this.

This is a very straightforward knot; watch the photo and follow the directions you see.

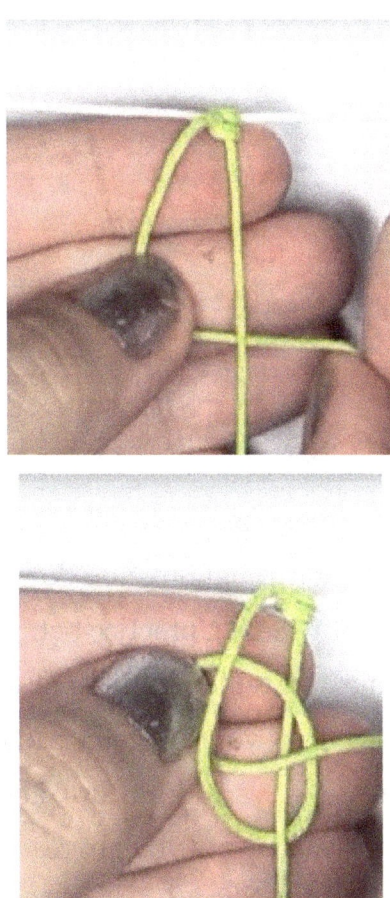

Pull the end of the cord up and through the center.

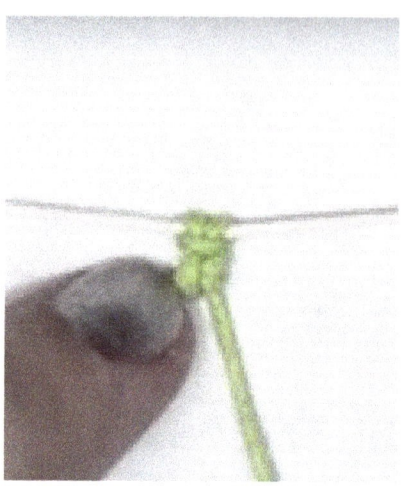

For the finished project, make sure that you have all your knots secure and firm throughout, and do your best to make sure it is all even. It will take practice before you can get it correctly each time, but remember that practice does make perfect, and with time, you will get it without too much trouble. Make sure all is even and secure, and tie off. Snip off all the loose ends, and you are ready to go!

# Horizontal Double Half Knot

This is an excellent beginning knot for any project and can be used as the foundation for its base. Use a lightweight cord for this. It can be purchased at craft stores or online, wherever you get your macramé supplies.

Oversee the photos as you move along with this project, and take your time to make sure you are using the right string at the right point of the project. Don't rush, and make sure you have even tension throughout. Practice makes perfect, but with the illustrations to help you, you'll find it's not hard at all to create.

Start at the top of the project and work your way toward the bottom. Keep it even as you work your way throughout the piece. Tie the knots at 4-inch intervals, working your way down the entire thing.

For the finished project, make sure that you have all your knots secure and firm throughout, and do your best to make sure it is all even. It will take practice before you can get it correctly each time, but remember that practice does make perfect, and with time, you will get it without too much trouble.

Make sure all is even and secure, and tie off. Snip off all the loose ends, and you are ready to go!

## Josephine Knot

This is the perfect knot to use for basket hangings, decorations, or any projects requiring you to put weight on the project. Use a heavier weight cord for this, which you can find at craft stores or online.

Oversee the photos as you move along with this project, and take your time to make sure you are using the right string at the right point of the project. Don't rush, and make sure you have even tension throughout. Practice makes perfect, but with the illustrations to help you, you'll find it's not hard at all to create.

Use the pins along with the knots that you are tying, and work with more extensive areas all at the same time. This will help you keep the project in place as you continue to work throughout the piece.

Pull the knots' ends through the loops, and form the ring in the strings' center.

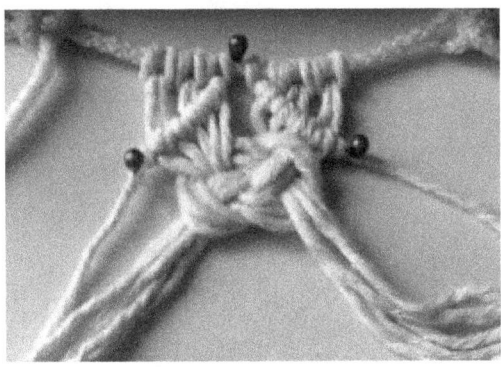

For the finished project, make sure that you have all your knots secure and firm throughout, and do your best to make sure it is all even. It will take practice before you can get it correctly each time, but remember that practice does make perfect, and with time, you will get it without too much trouble. Make sure all is even and secure, and tie off. Snip off all the loose ends, and you are ready to go!

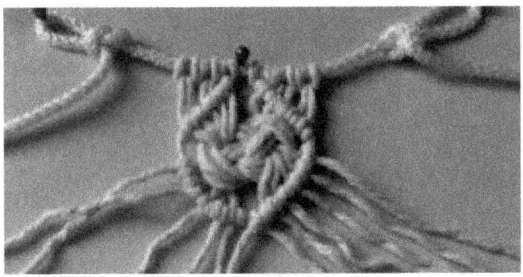

# Lark's Head Knot

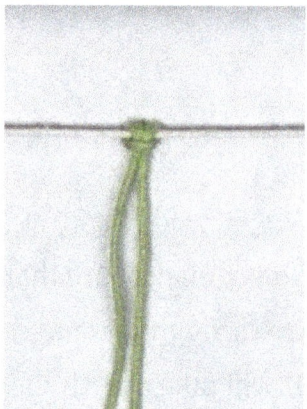

This is an excellent beginning knot for any project and can be used as the foundation for its base. Use a lightweight cord for this. It can be purchased at craft stores or online, wherever you get your macramé supplies.

Oversee the photos as you move along with this project, and take your time to make sure you are using the right string at the right point of the project. Don't rush, and make sure you have even tension throughout. Practice makes perfect, but with the illustrations to help you, you'll find it's not hard at all to create.

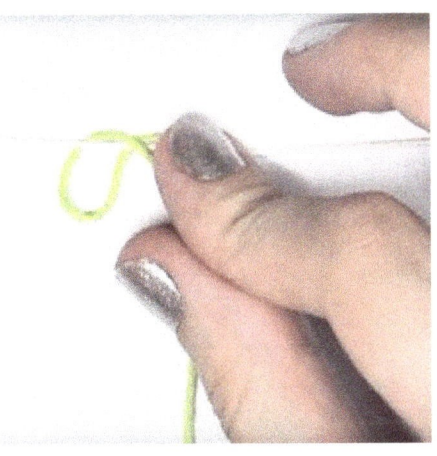

Use the base string as the core part of the knot, working around the string's end with the cord. Make sure all is even as you loop the string around the base of the cord.

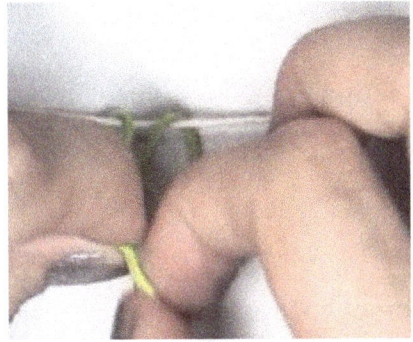

Create a slip knot around the string's base and keep both ends even as you pull the cord through the piece's center.

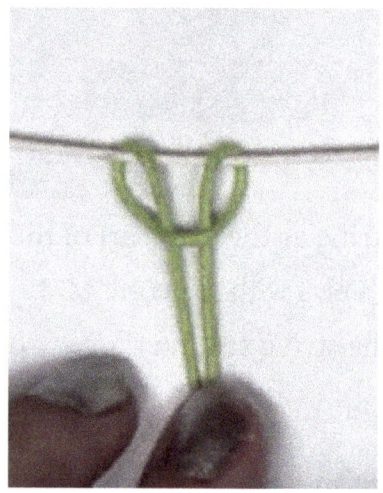

For the finished project, make sure that you have all your knots secure and firm throughout, and do your best to make sure it is all even. It will take practice before you can get it correctly each time, but remember that practice does make perfect, and with time, you will get it without too much trouble. Make sure all is even and secure, and tie off. Snip off all the loose ends, and you are ready to go!

# Reverse Lark's Head Knot

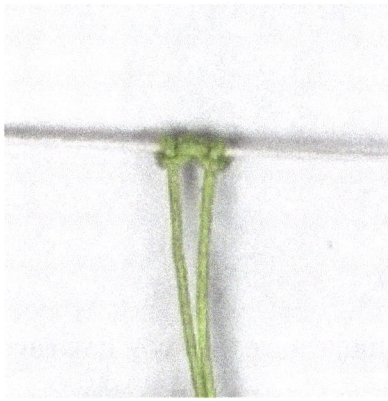

This is an excellent beginning knot for any project and can be used as the foundation for its base. Use a lightweight cord for this. It can be purchased at craft stores or online, wherever you get your macramé supplies.

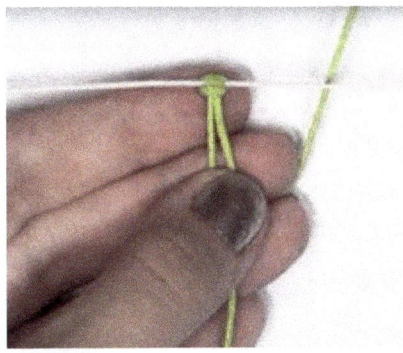

Oversee the photos as you move along with this project, and take your time to make sure you are using the right string at the right point of the project. Don't rush, and make sure you have even tension throughout. Practice makes perfect, but with the illustrations to help you, you'll find it's not hard at all to create.

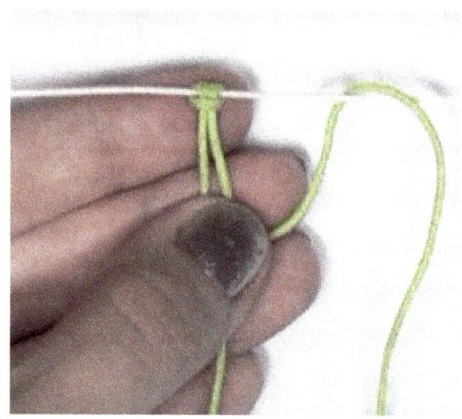

Use two hands to make sure that you have everything even and tight as you work. You can use tweezers if it helps to make it tight against the base of the string.

Use both hands to pull the string evenly down against the base string to create the knot.

Again, keep the base even as you pull the center, creating the firm Knot against your guide cord.

For the finished project, make sure that you have all your knots secure and firm throughout, and do your best to make sure it is all even. It will take practice before you can get it correctly each time, but remember that practice does make perfect, and with time, you will get it without too much trouble. Make sure all is even and secure, and tie off. Snip off all the loose ends, and you are ready to go!

## Lark's Head Half Stitches Knot

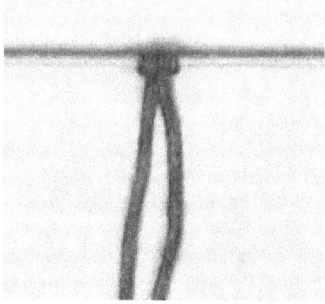

This is an excellent beginning knot for any project and can be used as the foundation for its base. Use a lightweight cord for this. It can be purchased at craft stores or online, wherever you get your macramé supplies.

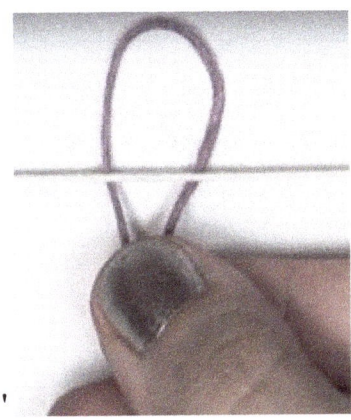

Oversee the photos as you move along with this project, and take your time to make sure you are using the right string at the right point of the project. Don't rush, and make sure you have even tension throughout. Practice makes perfect, but with the illustrations to help you, you'll find it's not hard at all to create.

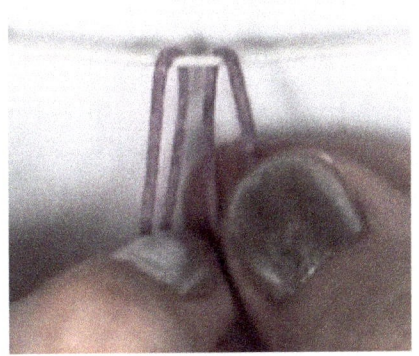

You will work this the same as the lark's head, just going in the opposite direction. Make sure you keep it firm against the cord's base and work through the steps as you did with the last. Watch the photos as a guide, following each step as you see them outlined there.

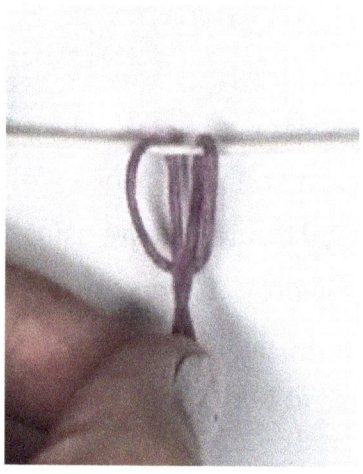

For the finished project, make sure that you have all your knots secure and firm throughout, and do your best to make sure it is all even. It will take practice before you can get it correctly each time, but remember that practice does make perfect, and with time, you will get it without too much trouble. Make sure all is even and secure, and tie off. Snip off all the loose ends, and you are ready to go!

# Single Half Stitch

This is an excellent beginning knot for any project and can be used as the foundation for its base. Use a lightweight cord for this. It can be purchased at craft stores or online, wherever you get your macramé supplies.

Observe the photos as you move along with this project, and take your time to make sure you are using the right string at the right point of the project. Don't rush, and make sure you have even tension throughout. Practice makes perfect, but with the illustrations to help you, you'll find it's not hard at all to create.

Use both hands to work around the cord, and make sure you follow each loop before you put on the next loop. One step at a time, as you see in the photo, and you're going to be okay.

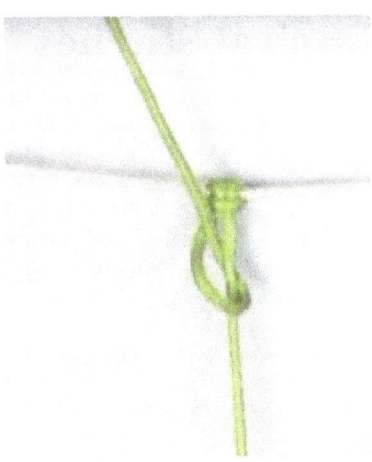

For the finished project, make sure that you have all your knots secure and firm throughout, and do your best to make sure it is all even. It will take practice before you can get it correctly each time, but remember that practice does make perfect, and with time, you will get it without too much trouble. Make sure all is even and secure, and tie off. Snip off all the loose ends, and you are ready to go!

# Spiral Stitches

This is the perfect knot to use for basket hangings, decorations, or any projects requiring you to put weight on the project. Use a heavier weight cord for this, which you can find at craft stores or online.

# Macramé Projects I

## DIY Macramé Feathers

Beautiful, wispy macramé feathers clogged my social media pages as late as possible, but I'm not crazy about it. They are lovely, and I certainly got to buy them, so I had to bookmark and hang them in the children's room. But I was also curious about how they were made, of course. How do you achieve this perfectly soft fringe in the world? It involves a brush of a cat. Enough has been said. Honestly, though, there are endless possibilities here, and I could not wait to play more with this technique. But I hope I'll inspire you to do these at home in the meantime.

Cut for a medium-sized feather:

- **1 32" strands around the spine**
- **10-12 14" strands around the top**
- **8-10 12" strands around the middle**
- **6-8 10" strands around the bottom**

Fold in half the 32 "strand. Choose one of the 14", fold them in half, and tuck them under the spine. Take the next 14 "strand, fold it in half and attach it to the top horizontal strand loop.

Pull this through and lay it horizontally on top of the oppositional strand. Now pull the bottom strands through the top loop all the way. Pull both ends of the spectrum tightly. In the next row, you will alternate the starting side. Pull through the top loop the lower threads, and squeeze. Keep going and work in scale slowly. Push up the strands to tighten-grab the lower end of the middle strand (spine) with one hand and push up the strands with the other. Once finished, drag the fringe down to meet the middle strand's bottom. Give her a rough trim instead. Not only does this help guide the shape, but it also helps to brush out the strands. To be honest, the shorter the lines, the better. It also helps to have a very sharp tissue shear pair! Start at the spine when brushing and push hard when brushing into the thread. To get that perfect, delicate fringe, it will take many hard strokes. Work down your way. Do not want the brush to throw off any strands while you brush the lower end of the spinal cord. You're going to want to stiffen the feather next. The cords are soft that it is just flopping if you collect it and try to hang it. Give it a spray or two, and allow at least a few hours to be tried.

# DIY Tassel and Macramé Keychains

Who doesn't love a sweet ring? Particularly a lovely DIY version which takes no time to make, uses stuff you've already got, can be as straightforward or as fantastic as you want? If you need an excuse to make a personalized keychain, we have you:

Update your keychain before remembering it, create a replacement set of keys for your domestic pet sitter, make a replacement set of keys you can leave from your neighbor so that when you lockout, you do not have to break into your place. Organize yourself by making a unique keyring for all those small rewards cards. Use this to improve your macramé skills. For the stripped Macramé keychain, I used vertical clove hitch knots and wool-roving thread. The third and fifth personalized keychain is super easy – strings and a few perles with a tassel. And the fourth DIY keychain is only a long braid, folded in half, wrapped in embroidery floss. Beads are hand-painted and made from Sculpey.

Materials needed for Macramé Keychains
- Key Ring
- 3/16" Natural Cotton Piping Cord
- Beads
- Embroidery yarn or floss
- Scissors

You can make things fancy on your keychain tassel or macramé by wrapping them in different yarn or floss colors.

## Simple Modern DIY Macramé Wall Hanging

Here's one of Macramé's funny thing 😊
They may think the gigantic macramé wall's hanging patterns are super complicated to complete, but trust me, they're not! It's an easy idea as long as you keep the knots and the overall design relatively simple. That's precisely what this project is about macramé.

The shape! It's extra-long and slim, and it's going to work in those itty-bitty random wall areas where you need some texture. Or hang it out of a door. It's a decent size. It's a pretty little nod for your decor.

Materials Needed for Macramé wall hanging

- **Macramé Rope**– I have been using this 4 mm rope– 12– 16′ (as in feet) cords are required (twelve). Note that this is a thick hanging wall, which is why we need longer cords. To act as your hanger, you will also need 1 shorter piece of cording. Simply tie it on either end with a simple knot.
- **A dowel or a stick**– I used a long (haha) knitting needle. As long as it is straight and robust and as long as you need to work with what you have!

**Here is the step-by-step guide for Macramé hanging wall.**

The first thing you want to do with each end is to knot a cord. For our project, this will serve as the hanger.

Making a macramé wall hanging when it's hanging is much easier than lying flat.

This can be hanged from cabinet knobs, doorknobs, a wreath hanger, or even a hanger for a picture. Just make sure it is robust!

1. Begin by folding in half your 16′ cords.
2. Make sure that the ends are the same.
3. Place the cord loop under the dowel and thread through the loop to the ends of the rope, and pull near.
4. That's the Head Knot of your first Reverse Lark. (For assistance, refer to basic macramé knots).
5. Repeat with the other 11 cords.
6. First, make 2 Square Knots rows.
7. Now make 2 rows of Square Knots Alternating.

8. Now make 2 more Square knots sets.
9. Follow this pattern until you have 10 rows in total (2 rows of knots in a square, 2 rows of square knots in alternation).
10. Working from left to right– make two half-hitch knots across your piece in a diagonal pattern.
11. Now, from right to left– make double half-hitch knots across your piece in a diagonal pattern.
12. Have 4 rows in all.
13. Make 2 more rows of knots of the square.
14. We will finish the hanging wall with a set of spiral knots.

This is a half-square node sequence (or left-side square branch). (Do not end on the right side of the knot, just make square knots and spiral on the left side for you again and again.) To build this spiral, I made a total of 13 half-square knots. Finally, I trimmed in a straight line the bottom cords. The total size for the hanging of my wall is 6.5″ wide by 34.

# DIY Round Macramé Boho Coasters

I don't seem to be able to rest every time I find a craft idea until I know how to do that; these coasters are the perfect example; I've done a macramé bracelet before, but to make a macramé round is strange for me, after exploring the internet, finding some confusing posts, and making my first coaster after ways to create it. That's the way I found it easier (although there's another one I tested as well), and it also got me the most beautiful result.

Supplies:
- **3 mm cotton cord**
- **Something to hold the cords, a cork coaster or a board, either tape or as I used it.**
- **Pins to hold if the cork is used.**
- **Fabric Scissors**
- **Ruler or measuring tape**
- **Comb**

**Boho Christmas Trees**

1. Cut the yarn into 7-8-inch bits.
2. Take two strands and fold them in half to form a loop.
3. Place one of the loops under a twig.
4. Start with the other strand's bowed end and move the strand's ends under the twig through the loop.
5. Thread through the loop below the twig the ends of that strand.

6. Pull tightly and repeat, okay? If you have added enough knotted strands, use a brush or a comb to separate the threads.
7. The "almost finished" tree will be a little floppy, so you have to stiffen it with some starch.
8. When erect, cut the Boho Christmas trees into a triangle shape and decorate them with small baubles or beads.

I just made the jewelry wire a little flower star. They're going to take about 10 minutes to make a whole bunch. I think they'd make beautiful gifts or you could hang them on your Christmas tree.

## Macramé Curtain

1. Tie four strands together on the same foam core board and place pins in the top knot to keep those in place beneath the two center strands.
2. Take a right outer strand (pink) and pass it over the other two center fibers to the left side.
3. Take a left (yellow) outer strand and pass it under the pink strand behind the middle fibers and on the other side over the pink strand.
4. Push the two strands near.
5. Then, in the first step, you just reverse what you did!
6. Take the left-most strand (now the pink) and lay it over two strands in the middle.

7. Take the extreme right strand (which is now the yellow) and transfer it under the pink, behind the two middle strands, and on the other side over the pink.
8. Pull the two strands tightly until they create a knot from the woven strands.
9. That's the most challenging part! The rest of the steps repeats these basic motions.
10. To make another knot right next to your first knot, repeat steps 1-3 with four more threads.
11. Put in two right strands of the first knot for a new group with two left-most strands of the second knot.
12. Repeat with the new group your basic knot by taking the outer right (purple) strand and passing it over the middle two strands to the left side.
13. Take the outside (green) left and pass it through the purple strand, after the medium strands, and across the purple strand on the other side.
14. Push the two fibers near.
15. Now turn the first step back! Take the left-most strand (purple now) and lay it over two strands in the middle.
16. Take the extreme right strand (the green) and pass it under the purple, behind the two middle strands, and across the violet on the other side.
17. Pull these two threads near.
18. By moving the two left-most threads and the two rightmost strands, divide the middle group of strands.
19. Repeat the fundamental knot with both classes and continue until so long as you like.

I built 14 rope classes, each with four lines, all 100 inches long, when I began the actual curtain. It made a clean knot to cut two cords twice as long (thus 200 inches) on top of the curtain and then hung strands across the rod at the middle point and tied up a knot to create a group of four strands.

Since doing this method with big ropes is much larger than the thread, you're going to have to find something to hang your rod from, so you let your rope hang under it (we've used a bike rack to hang our rope). You can see that making the superficial knots with the yarn is the same idea, but only on a much larger scale. I rendered the base knot near the top of all 14 bands and then made a new knot line below and above them (as in the yarn instructions). I then went down a new row, took knots under the original knots, and kept the knots rowing until I had finished all the rows I needed. Make sure you stay back while you make your knots to make sure you tie your knots in even rows. I maintained a handy ruler to measure each knot's distance and the wooden curtain rod to make sure that they turned out. I dropped the rest of the strands to complete the curtain until 5 rows of knots were completed.

Hang your new curtain in your perfect place once you have finished braiding the ropes. Finally, wrap the masking cassette at the ends where the cloth reaches the ground, 6 1/2 feet tall (or white cassette, which is "dorm cassette') I have used. Slice the candle and leave intact 2/3 to half of the candle. It helps to prevent the spawning of overtime. I hung an off-white piece of textile on the macramé curtain (on the existing clothes rack). I love how it turned out that curtain! It feels different but still functional, not too loud. We've got a very loud rug in the room, so we didn't need anything with tons of color or much attention.

## Friendship Bracelet Watch

You'll need your watch face and floss to get started. I use art floss in the colors of brown, white, and minty blue. Cut strips approximately 48 inches long. You will need 10 of these long strands for each side for this watch face (but just cut 10 right now, leave the others until you're ready to start from the other side). To start making our harness, we will lash each piece of floss onto the chain. Bring together the ends of a long piece of floss and pick up the end. Push through the bar and pull the ends through the loop you've built. Start with all your floss cutting. Make sure you keep the colors in your pattern as you want them. I wanted thick orange and mint stripes and thin white stripes.

My order was, therefore: orange, orange, white, mint, mint, mint, mint, mint, white, orange, orange.

And now you're just starting to braid your friendship. We won't have this weird thing bundled up in most friendship bracelets that begin with a knot because we've latched on to the message. Pretty better, huh? Like any other friendship bracelet, you have the choice to twist and then tie when you are wearing it. This is not the most beautiful option, but it's going to work well. But if you want to use closures, continue reading. You want to take a decent amount of glue when you get the length and run a line where you need to cut. Apply the glue on the front and back sides of the threads. For our next move, this will hold the braid securely together. I used the fast-dry tacky glue from Aleena because I'm very impatient. I didn't think about this, and I had to shorten my straps after a few wears. Perhaps you'd like to go ahead and make the watch a bit tight. The first wear may be uncomfortable, but it will be perfect for a couple of hours.

Use sharp scissors to cut the area where you applied the glue through your strap. See how well it sticks with each other? Go ahead and run a little at the very end to help avoid fraying. Place the clamp on the end of the straps and use the pin to lock on firmly. Finish with a jumping ring on one and a jumping ring and closing on the other. And you got it there! It's a pretty fun wear watch and brings the friendship bracelet's whole trend in a new way. What do you think about it? Are you going to make one? It sounds like an excellent project for me at the weekend!

# Macramé Projects II

## Serenity Bracelet

(Note: if you are familiar with the flat knot, you can move right along into the next pattern)

This novice bracelet offers plenty of practice using one of the micro macramé's most used knots. You will also gain experience in beading and equalizing tension. This bracelet features a button closure, and the finished length is 7 inches.

Knots Used:

- **Flat knot (aka square knot)**
- **Overhand Knot**

Supplies:

- **White C-Lon cord, 6 ½ ft, x 3**
- **18 - Frosted Purple size 6 beads**
- **36 - Purple seed beads, size 11**
- **1 - 1 cm Purple and white focal bead**
- **26 - Dark Purple size 6 beads**

- **1 - 5 mm Purple button closure bead**

(Note: the button bead needs to be able to fit onto all 6 cords)

Instructions:
1. Take all 3 cords and fold them in half.
2. Find the center and place it on your work surface as shown:

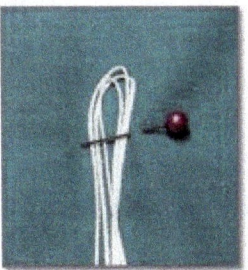

3. Now hold the cords and tie an overhand knot, loosely, at the center point.
It should look like this:

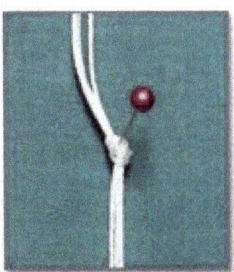

## We will now make a buttonhole closure.

1. Just below the Knot, take each outer cord and tie a flat knot (aka square knot). Continue tying flat knots until you have about 2 ½ cm.

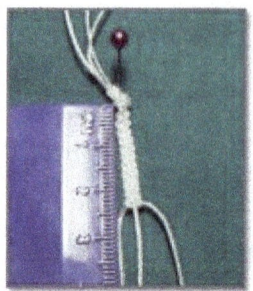

2. Undo your overhand knot and place the ends together in a horseshoe shape.

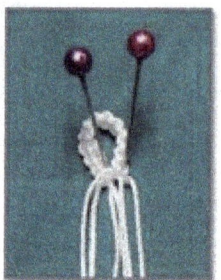

3. We now have all 6 cords together. Think of the cords as numbered 1 through 6 from left to right. Cords 2-5 will stay in the middle as filler cords. Find cord 1 and 6 and use these to tie flat knots around the filler cords. (Note: now, you can pass your button bead through the opening to ensure a good fit. Add or subtract flat knots as needed to create a snug fit. This size should be sufficient for a 5mm bead). Continue to tie flat knots until you have 4 cm worth. (To increase bracelet length, add more flat knots here, and the equal amount in step 10).

4. Separate cords 1-4-1. Find the center 2 cords. Thread a size 6 frosted purple bead onto them, then tie a flat knot with cords 2 and 5.

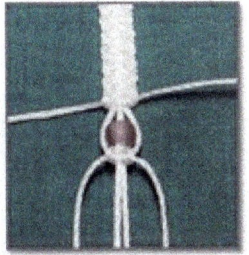

5. We will now work with cords 1 and 6. With cord 1, thread on a seed bead, a dark purple size 6 bead, and another seed bead. Repeat with cord 6, then separate the cords into 3-3. Tie a flat knot with the left 3 cords. Tie a flat knot with the right 3 cords.

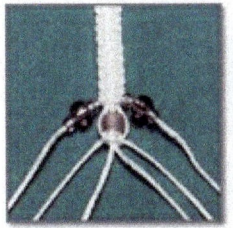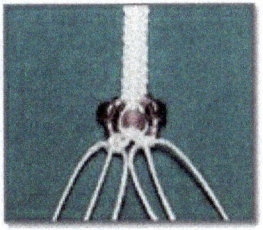

6. Repeat steps 4 and 5 three times.

7. Find the center 2 cords, hold them together, and thread on the 1cm focal bead. Take the next cords out (2 and 5) and bead as follows: 2 sizes 6 dark purple beads, a frosted purple bead, 2 dark purple beads. Find cords 1 and 6 and bead as follows: 2 frosted purple beads, a seed bead, a dark purple bead, a seed bead, 2 frosted purple beads.

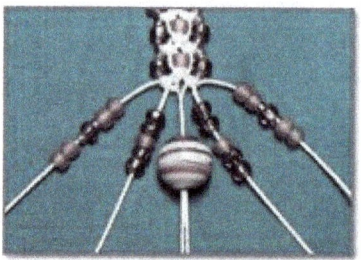

8. With cords 2 and 5, tie a flat knot around the center 2 cords. Place the center 4 cords together and tie a flat knot around them with outer cords 1 and 6.

9. Repeat steps 4 and 5 four times.

10. Repeat step 3.

11. Place your button bead on all 6 cords and tie an overhand knot tight against the bead. Glue well and trim the cords.

# Lantern Bracelet

This pattern may look simple, but please don't try it if you are in a hurry. This one takes patience. Don't worry about getting your picot knots all the same shape. Have fun with it! The finished bracelet is 7 ¼ inches in length. If desired, add a picot knot and a spiral knot on each side of the centerpiece to lengthen it. This pattern has a jump ring closure.

Knots Used:

- **Lark's Head Knot**
- **Spiral Knot**
- **Picot Knot**
- **Overhand Knot**

Supplies:

- **3 strands of C-Lon cord (2 light brown and 1 medium brown) 63-inch lengths**
- **Fasteners (1 jump ring, 1 spring ring, or lobster clasp)**
- **Glue - Beacon 527 multi-use**
- **8 small beads (about 4mm) amber to gold colors**
- **30 gold seed beads**

- **3 beads (about 6 mm) amber color (mine are rectangular, but round or oval will work wonderfully also)**

Note: Bead size can vary slightly. Just be sure all beads you choose will slide onto 2 cords (except seed beads).

Instructions:

1. Find the center of your cord and attach it to the jump ring with a lark's head knot. Repeat with the 2 remaining strands. If you want the 2-tone effect, be sure your second color is NOT placed in the center, or it will only be a filler cord, and you will end up with a 1 tone bracelet.

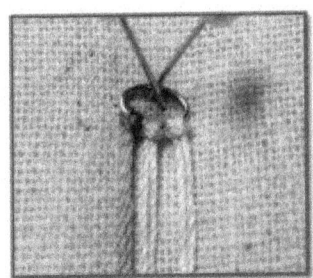

2. You now have 6 cords to work with. Think of them as numbers 1 through 6, from left to right. Move cords 1 and 6 apart from the rest. You will use these to work the spiral knot; all others are filler cords. Take cord number 1 to tie a spiral knot. Always begin with the left cord. Tie 7 more spirals.

3. Place a 4mm bead on the center 2 cords. Leave cords 1 and 6 alone for now and work 1 flat knot using cords 2 and 5.

4. Now, put cords 2 and 5 together with the center strands. Use 1 and 6 to tie a picot flat knot. If you don't like the look of your picot knot, loosen it up and try again. Gently tug the cords into place, then lock in tightly with the next spiral knot.

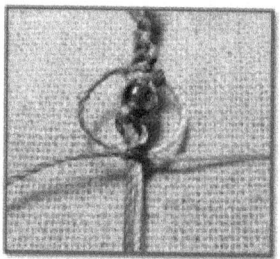

Notice how I hold the picot knot with my thumbs while pulling the cords tight with my fingers. If you look closely, you may be able to see that I have a cord in each hand.

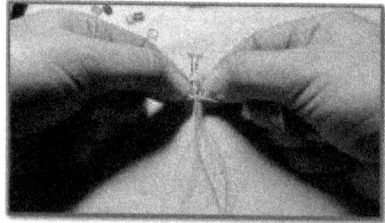

5. Tie 8 spiral knots (using left cord throughout pattern).

6. Place a 4mm bead on the center 2 cords. Leave cords 1 and 6 alone for now and work 1 flat knot using cords 2 and 5. Now put cords 2 and 5 together with the center strands. Use strands 1 and 6 to tie a picot flat knot.

7. Repeat steps 5 and 6 until you have 5 sets of spirals.

8. Next, place 5 seed beads on cords 1 and 6. Put cords 3 and 4 together and string on a 6 mm bead. Tie one flat knot with the outermost cords.

Repeat this step two more times.

Now repeat steps 5 and 6 until you have 5 sets of spirals from the center point. Thread on your clasp. Tie an overhand knot with each cord, glue well, and let dry completely. As this is the weakest point in the design, I advise trimming the excess cords and gluing again. Let dry.

## Celtic Choker

Elegant loops allow the emerald and silver beads to stand out, making this a striking piece. The finished length is 12 inches. Be sure to use the ribbon clasp, which gives multiple length options to the closure.

Knots Used:

- **Lark's Head Knot**
- **Alternating Lark's Head Chain**

Supplies:

- **3 strands of black C-Lon cord; two 7ft cords, one 4ft cord**
- **18 - green beads (4mm)**
- **7 - round silver beads (10 mm)**
- **Fasteners: Ribbon Clasps, silver**
- **Glue - Beacon 527 multi-use**

Note: Bead size can vary slightly. Just be sure all beads you choose will slide onto 2 cords.

Instructions:

1. Optional – Find the center of your cord and attach it to the top of the ribbon clasp with a lark's head knot. I found it easier to thread the loose ends through and pull them down until my loop was near the opening, then push the cords through the loop. Repeat with the 2 remaining strands, putting the four-foot cord in the center. If this is problematic, you could cut all the cords to 7ft and not worry about placement. (If you trust your glue, you can skip this step by gluing the cords into the clasp and going from there).

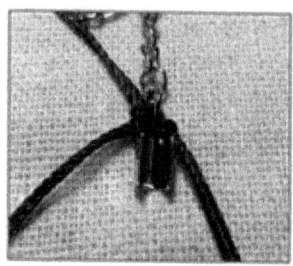

2. Lay all cords into the ribbon clasp. Add a generous dap of glue and use pliers to close the clasp.

3. You now have 6 cords to work with. Find the 4 ft cords and place them in the center. They will be the holding (or filler) cords throughout.

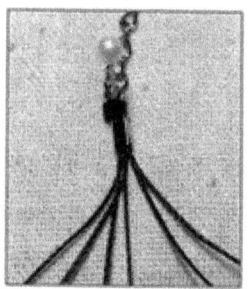

4. Begin your Alternating Lark's Head (ALH) chain, using the outmost right cord, then the outermost left cord. Follow with the other right cord than the last left cord. For this first set, the pattern will be hard to see. You may need to tug gently on the cords to get a little slack in them.

5. Now slide a silver bead onto the center 2 cords.

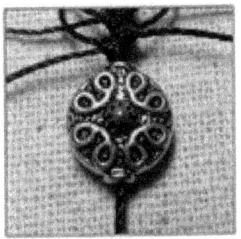

6. The outer cords are now staggered on your holding cords. Continue with the ALH chain by knotting with the upper right cord...

then tie a knot with the upper left cord.

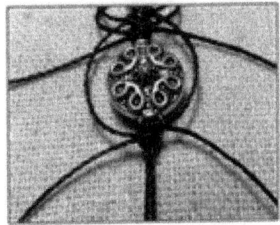

7. Finish your set of 4 knots, then add a green bead

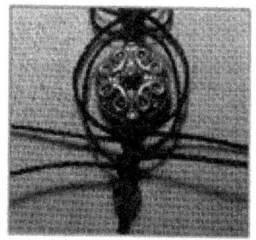

8. Tie four ALH knots followed by a green bead until you have 3 green beads in the pattern. Then tie one more set of 4 ALH knots.

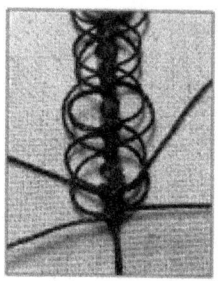

9. Slide on a silver bead and continue creating sequences of 3 green, 1 silver (always with 4 ALH knots between each). end with the 7th silver bead, and 1 more set of 4 ALH knots, for a 12" necklace.

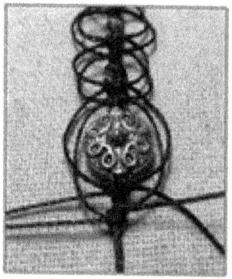

10. Lay all cords in the ribbon clasp and glue well.

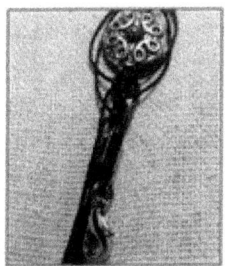

11. Crimp shut, let dry completely and trim excess cords.

# Climbing Vine Keychain

This pattern is a fun way to practice the Diagonal Double Half-Hitch knot. It works up quickly and is a fun piece to work in various colors. Just be sure to use enough beads on the fringe work to weigh the threads down.

Knots Used:
- **Lark's Head**
- **Flat Knot**
- **Diagonal Double Half-Hitch**

Supplies:
- Measure out 3 cords of Peridot C-Lon, 30" each
- 1 key ring
- 2 (5mm) beads
- 8 (plus extra for ends) pink seed beads
- 4 (plus extra for ends) gold seed beads
- 12 (plus extra for ends) green seed beads
- 8 (plus extra for ends) 3mm pearl beads (seed pearl beads will work also)

- **Glue - Beacon 527 multi-use**

Note: You can vary slightly to the bead size. Just be sure that 2 cords will fit through the 2 main beads (the 5mm size beads)

Instructions:

1. Fold each cord in half and use a lark's head knot to attach it to the key ring. Secure onto your work surface with straight pins. You now have 6 cords to work with.

2. Separate cords into 3 and 3. Using the left 3 cords, tie 2 flat knots. Repeat with the right 3 cords.

3. Place all six cords together and think of them as numbered 1-6, left to right. Skip cord 1 and place a pink seed bead on cord 2. Skip cord 3 and place 2 gold seed beads on cord 4. Skip cord 5 and put 3 pink seed beads on cord 6.

4. Using cord 1 as your holding cord, tie a row of diagonal double half-hitch (DDHH) knots beginning on the left and ending on the right. Using cord 1 on the left, move it to the right as a holding cord and tie DDHH knots to the right.

5. Put all six cords together and place 7 small beads on cord 1. Skip cord 2 and string your focal bead onto cords 3 and 4. Skip cord 5 and put 3 small beads on cord 6.

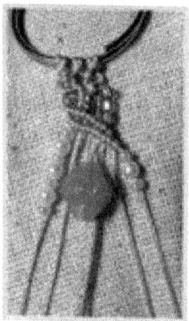

6. Use cord 6 as your holding cord and tie a row of DDHH knots from right to left. Repeat once more.

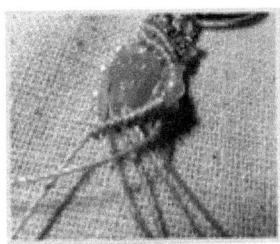

7. Repeat beading from step 3.

8. Repeat a row of diagonal double half-hitch knots from step 4 (left to right) twice.

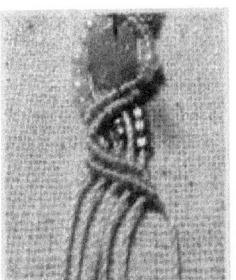

9. Bead as stated in step 5.

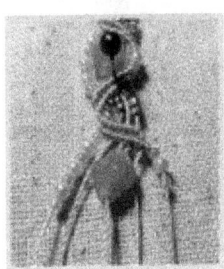

10. Repeat a row of DDHH knots as written in step 6 (right to left) twice.

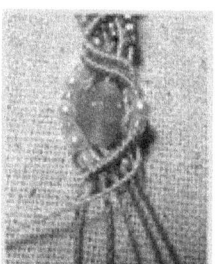

11. Separate cords into 3 and 3. Tie 1 flat knot with the left 3 cords and 2 flat knots with the right 3 cords.
12. Separate cords into 1 – 4 – 1 and tie 1 flat knot with the center 4 cords only, letting cords 1 and 6 float.
13. Separate cords into 3 and 3. Tie 1 flat knot with each section.
14. Repeat step 12.

15. Bead ends with various size beads. Be sure there is enough weight to hold the ends downward. Tie an overhand knot with each cord and glue well. Let dry completely and trim cords.

# Filigree Lancelet Bracelet

The Overhand Knot, Flat Knot, Alternating Lark's Head knot, and Diagonal Double Half Hitch knot are all in play. This "Lace-let" fits the very definition of filigree as it is both delicate and fanciful.

**I hope you enjoy this open design and light.**

The finished length is 7 1/2 inches and includes a button closure.

Knots Used:
- **Overhand Knot**
- **Diagonal Double Half Hitch**
- **Flat Knot**
- **Alternating Lark's Head Knot**

Supplies:
- **66" length white C-Lon cord, 4 strands**
- **6 clear beads, 5mm**
- **56 clear beads, 3mm**
- **5 clear beads, 4mm**
- **1 bead for a button closure, about 7mm**
- **164 clear seed beads**
- **Glue - Beacon 527 multi-use**

Note: You can vary the bead sizes slightly. Just be sure the beads you choose will slide onto 2 and sometimes 3 cords. (The seed beads only need to fit onto one cord).

Instructions:

1. Pin this onto your project board. Tie about 9 flat knots (for 7mm button closure bead). Now undo the overhand knot and fold the flat knots into a horseshoe shape. Using the outer cord from each side, tie 1 flat knot.

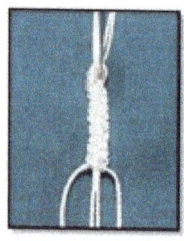 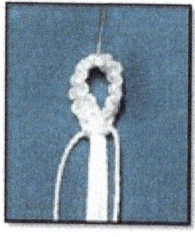

2. Take the rightmost cord and place it over all others down to the left to work Diagonal Double Half Hitch (DDHH) knots from right to left. Put 1 clear seed bead on each cord tie another set of DDHH knots from right to left.

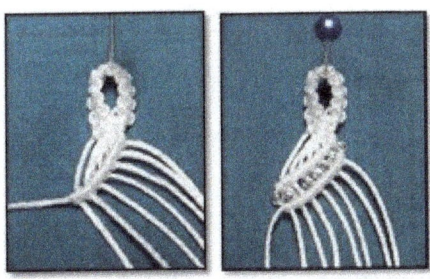

3. Separate cords into 4-4. Working with left 4 cords bead as follows: On the left-most, cord put 4 clear 3mm beads with a seed bead between each one. The next cord gets 5 clear seed beads. The next cord needs a 5mm clear bead. And the last cord of this section gets 5 clear seed beads. Use the outer 2 cords to tie a flat knot around the inner cords.

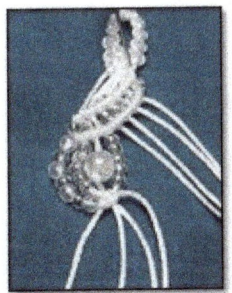

4. Working with right 4 cords: Place a 3mm clear bead on the center 2 cords. Place a seed bead on the rightmost cord. Now use this rightmost cord to tie an Alternating Lark's Head (ALH) knot around the other 3 cords. Repeat 4 times.

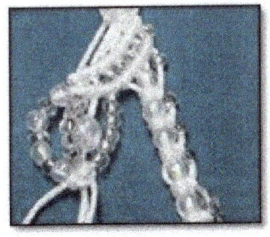

5. Using the left-most cord as a holding cord, work DDHH knots from left to right. Place a seed bead on each cord, then work another set of DDHH knots (from left to right again) using the left-most cord as your holding cord.

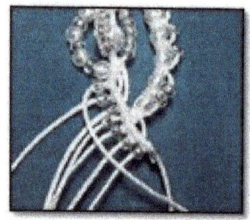 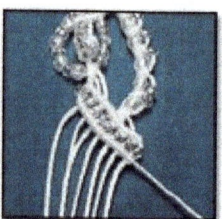

6. Separate cords into 4-4. Working with left 4 cords: Place a 3mm clear bead on the center 2 cords. Place a seed bead on the left-most cord. Now use this left-most cord to tie an ALH knot around the other 3 cords. Repeat 4 times.

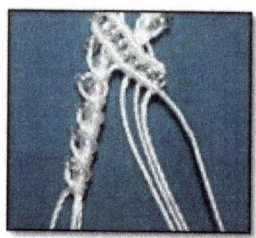

7. Working with the right 4 cords: the rightmost cord gets 4 clear 3mm beads with a seed bead between each one. The next cord in from the right needs 5 seed beads. The next cord gets a 5mm clear bead. And the last cord of this section gets 5 seed beads. Use the outer 2 cords to tie a flat knot around the inner cords.

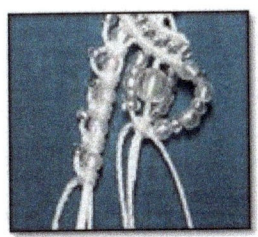

8. Repeat steps 2-7 for pattern until you have about 6 1/2 inches in length.

9. Separate cords into 3-2-3. On the left set of cords, place a 4mm bead. With the center 2 cords thread on a 3mm bead, a 4mm bead, and another 3mm bead. On the right, 3 cords place three 4mm beads. Find the outermost cord on each side and tie a flat knot around the rest.

10. Thread your button bead onto the center 4 or 6 cords if possible. Use the outer cords to tie a flat knot. Glue flat knot, let dry and trim excess cords.

# Instruction Books

Learn macramé with step-by-step instructions accompanied by real knot pictures.

# Plant Hanger Ayla

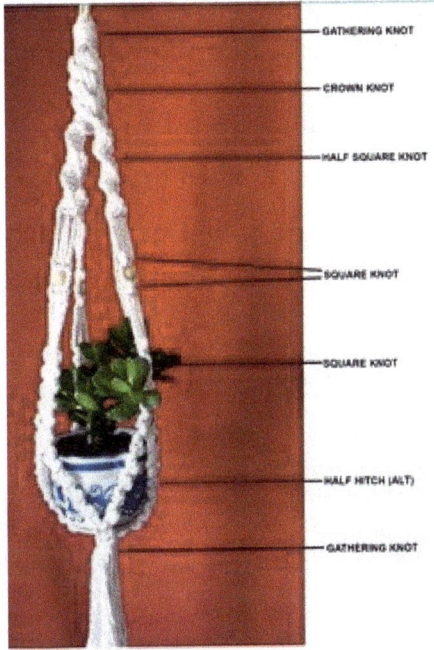

Description: Plant hanger of 2 feet and 3,5 inches (70 cm)

Used Knots:

- **Square Knot**
- **Half square knot**
- **Alternating Square Knot**
- **Crown knot**
- **Gathering knot**
- **Half hitch knot**

Supplies:

- **4 strands of a cord of 13 feet and 1,5 inches (4 meters)**

- 4 strands of 16 feet and 4,8 inches (5 meters)
- 2 strands of 3 feet and 3,4 inches (1 meter)
- 1 wooden ring of 2 inches (50 mm)
- 4 wooden beads: diameter 0,4 inches (10mm)

Directions (step-by-step):

1. Fold the 8 longer strands of cord in half through the wooden ring. Tie all (now 16) strands together with 1 shorter strand of 3 feet and 3,4 inches (1 m) with a gathering knot. Cut the cord ends off after tying the gathering knot.

2. Now follows the crown knot. It is easiest to turn your project upside-down between your legs, as shown in the photos. Divide the 16 strands into 4 sets of 4 strands each. Each set has 2 long strands and 2 shorter strands. Tie 5 crown knots in each set. Pull each strand tight and smooth.

3. Tie 15 half square knots on each set of four strands. In each set, the 2 shorter strands are in the middle, and you are tying with the 2 outer, longer strands. Dropdown 2,4 inches (6 cm of no knots).

4. Tie 1 square knot with each set.

5. Then add the wooden bead to the 2 inner cords and tie 1 square knot with each set again. Dropdown 2,4 inches (6 cm of no knots) and tie 6 square knots with each of the 4 sets.

6. Take 2 strands of 1 set and make 10 alternating half hitch knots. Repeat for the 2 left strands of that set, and then repeat for all sets.

7. Tie an alternating square knot to connect the left two cords in each set with the right two of the set next to it. This is followed by 3 square knots for each new set (so you have 4 square knots in total for each new-formed set).

8. Place your chosen container/bowl into the hanger to make sure it will fit, gather all strands together and then tie a gathering knot with the leftover shorter strand of 3 feet and 3,4 inches (1 m). Trim all strands to the length that you want. If you want, you can unravel the ends of each strand.

# Plant Hanger Bella

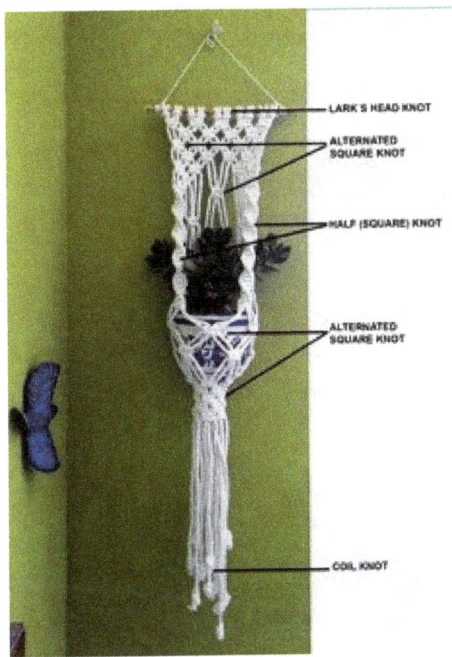

Description: Plant hanger of 60 cm (not counting the fringe)

Supplies:

- **6 strands of the cord of 13 feet and 1,5 inches (4 meters)**
- **4 strands of 16 feet and 4,8 inches (5 meters)**
- **Wooden stick of 11,8 inches (30cm)**

Used Knots:

- **Half knot**
- **Lark's Head knot**
- **(Alternating) Square Knot**
- **Coil knot**

Directions (step-by-step):
1. Fold all strands in half and tie them to the wooden stick with Lark's Head knot. The longest strands are on the outer side (2 strands on the left side and 2 at the right).

2. Make 4 rows of alternating square knots (see knot guide for explanation).

3. In the 5th row, you only make 2 alternating square knots on the right and 2 on the left.

4. In the 6th row, you only tie 1 alternating square on each side.

5. Then, with the 4 strands on the side, you tie 25 half (square) knots. Do this for both sides, the left and right sides.

6. Take 4 strands from the middle of the plant hanger, first drop down 2,4 inches (6 cm of no knots) and then tie a square knot with the 4 center strands. The 4 strands next to the middle dropdown 3,15 inches (8 cm of no knots) and tie a square knot. Do this for both sides (left and right).

7. Dropdown 2,4 inches (6 cm of no knots) and tie 2 (alternated) square knots by taking 2 strands from both sides (right and left group). Then 3 alternating square knots with the other groups. These knots must be about at the same height where the strands with the half knots have ended.

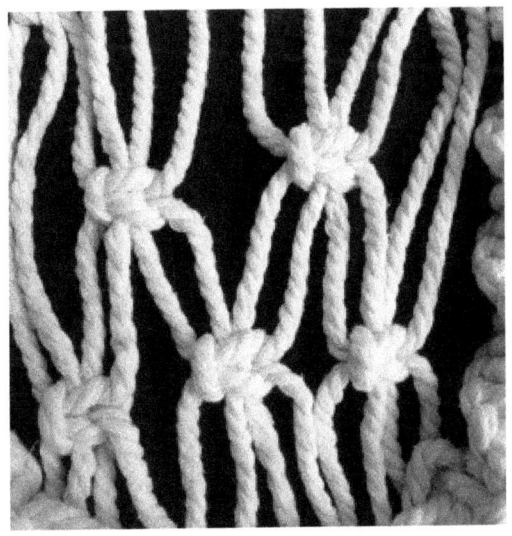

8. Take the 2 outer strands of the left group, which you made 25 half knots, and take the 2 outer strands of the group on the right. First, dropping down 2,4 inches (6 cm of no knots), you tie a square knot with these 4 strands.

9. Do the same with the rest of the strands leftover, make groups of 4 strands and tie alternated square knots on the same height as the one you made in step 8. Dropdown 2,4 inches (6 cm of no knots) and make another row of alternated square knots using all strands.

10. Dropdown 2,4 inches (6 cm of no knots) and make 5 rows of alternated square knots. Be careful: this time, leave NO space between the alternated square knots, and you make them as tight as possible.

11. Dropdown as many inches/cm as you want to make the fringe and tie at all ends a coiled knot.

12. Then cut off all strands directly under each coil knot.

# Plant Hanger Cathy

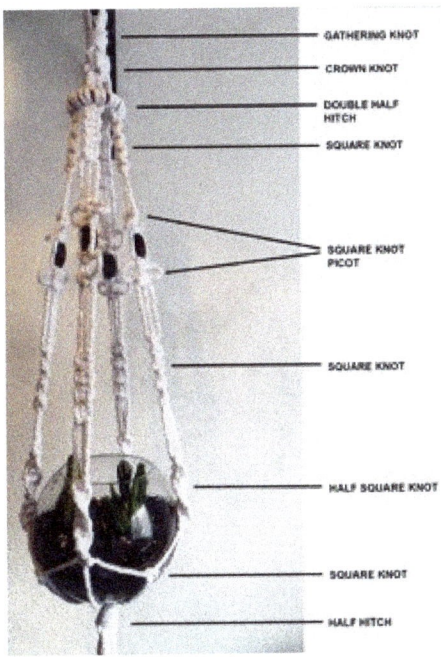

Description: Plant hanger of 2 feet and 9,5 inches (85 cm) - not counting the fringe

Supplies:
- 4 wooden beads of 1,2 inches (3cm)
- 3 inches (7,5cm) wooden ring
- 4 cords of 18 feet (5,5 meter)
- 2 cords of 15 feet (4,5 meter)
- 1 cord of 2 feet and 1,6 inches (65 centimeters)

Used Knots:
- **Gathering knot**
- **Crown knot**

- **(double) Half hitch**
- **(Half) Square Knot**
- **Square Knot**

Directions (step-by-step):

1. Fold the 6 longer cords in half, placing the loops neatly side by side. Use a gathering knot for tying the cords together with the shortest cord. This gives you twelve cords in total.

2. Arrange the cords in four groups of three cords each. Make sure that each group consists out of 2 longer cords and 1 shorter cord. Tie three Chinese Crown knots with the four groups of cords.

3. Slip the wooden ring over the top loop and drop it down 1,2 inches (3 cm) from the last Chinese Crown knot. With each of the twelve cords, tie one double half hitch on the ring to secure it. This gives you a ring of double half hitches.

4. Arrange the cords into four groups of three cords each. The middle cord of each group is the shorter one; this is called the filler cord. Repeat step five thru eight for each group.
5. Tie four square knots, each having one shorter filler cord.

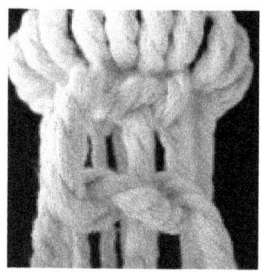

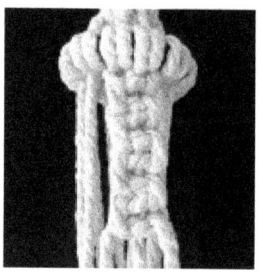

6. Skip down 2 inches (5 cm). Tie one square knot picot.

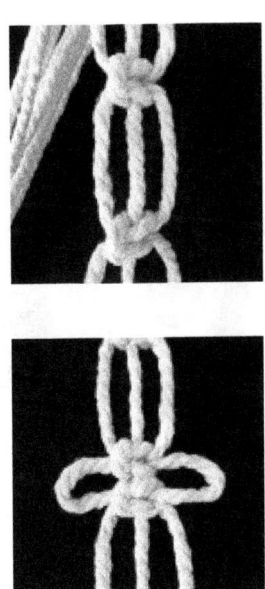

7. Slide a bead up the filler cord. Tie another square knot picot directly under the bead.

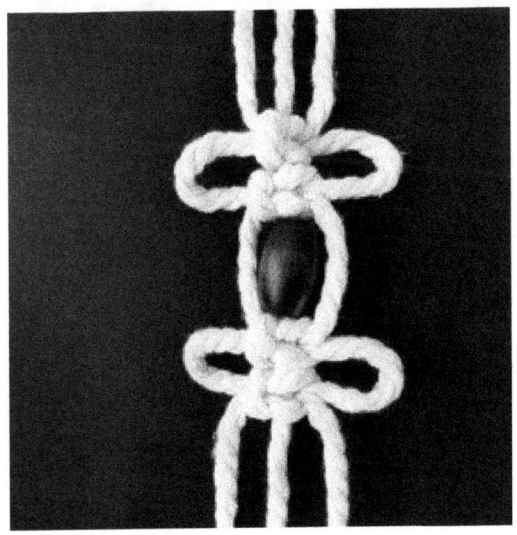

8. Skip down 2 inches (5 cm). Tie five square knots, each having one filler cord.

9. Skip down 2 inches (5 cm). Tie 10 half square knots, each having one filler cord.

10. Repeat the following procedure for each of the four groups you have just knotted: skip down 2,4 inches (6 cm); take one cord from each neighboring square knot to tie a square knot WITHOUT a filler cord. This gives you four square knots made of two cords each. The cords in the middle of each group are NOT used to knot.

11. Skip down 4,8 inches (12 cm). Gather and tie all cords together with one of the cords hanging using to tie 10 times a half hitch.

12. Cut the fringe to measure 6 inches (15 cm).

# Conclusion

The beauty of Macramé as a vintage art that has survived extinction for centuries and has continued to thrive as a technique of choice for making sophisticated but straightforward items merely is unrivaled. The simple fact that you have decided to read this manual means that you are well on your way to making something great. There is indeed a certain, unequaled feeling of satisfaction that comes from crafting your masterpiece.

The most crucial rule in Macramé is the maxim: "Practice makes perfect." If you cease to practice regularly, your skills are likely to deteriorate over time. So, keep your skills sharp, exercise the creative parts of your brain, and keep creating mind-blowing handmade masterpieces. Jewelry and fashion accessories made with even the most basic Macramé knots are always a beauty to behold. Hence they serve as perfect gifts for loved ones on special occasions. Presenting a Macramé bracelet to someone, for instance, passes the message that you didn't just remember to get them a gift, you also treasure them so much that you chose to invest your time into crafting something unique especially for them too, and trust me, that is a compelling message. However, the most beautiful thing about Macramé is that it helps create durable items. Hence you can keep a piece of decoration, or a fashion accessory you made for yourself for many years, enjoy the value, and still feel nostalgic anytime you remember when you made it. It even feels better when you made that item with someone. This feature of durability also makes Macramé accessories incredibly perfect gifts.

Macramé can also serve as an avenue for you to begin your small dream business. After perfecting your Macramé skills, you can conveniently sell your items and get paid well for your products, especially if you can correctly make items like bracelets that people buy a lot. You could even train people and start your own little company that makes bespoke Macramé fashion accessories. The opportunities that Macramé presents are genuinely endless.

So, stay sharp, keep practicing, and keep getting better. Welcome to a world of infinite possibilities!

# MACRAME

CPSIA information can be obtained
at www.ICGtesting.com
Printed in the USA
BVHW052247070421
604338BV00007B/645